American Agriculture:

AN

ADDRESS

DELIVERED BEFORE THE

BRISTOL COUNTY AGRICULTURAL SOCIETY,

ON OCCASION OF THEIR

ANNUAL CATTLE SHOW AND FAIR AT TAUNTON,

Oct. 15, 1852.

BY ROBERT C. WINTHROP.

BOSTON:
PRINTED BY JOHN WILSON & SON,
22, School Street.
1853.

American Agriculture:

AN

ADDRESS

DELIVERED BEFORE THE

BRISTOL COUNTY AGRICULTURAL SOCIETY,

ON OCCASION OF THEIR

ANNUAL CATTLE SHOW AND FAIR AT TAUNTON,

Oct. 15, 1852.

BY ROBERT C. WINTHROP.

BOSTON:
PRINTED BY JOHN WILSON & SON,
22, School Street.
1853.

At a meeting of the BRISTOL COUNTY AGRICULTURAL SOCIETY, Oct. 15, 1852, it was unanimously —

Voted, That the thanks of this Society be presented to Hon. Robert C. Winthrop for his eloquent and instructive address, and that Mr. Winthrop be invited to furnish a copy for the press.

ADDRESS.

I AM not insensible, Mr. President and Gentlemen of the Bristol County Agricultural Society, how adventurous a thing it is for one who has had so little personal acquaintance with agriculture as myself — for one who was born and brought up in a city of paved streets, in which it is our special boast that not a blade of grass is ever permitted to grow — to undertake a formal address to a society of practical farmers.

There are those within hearing who know, however, — and none better than yourself, sir, — that I am no volunteer on this occasion and in this service; that I am not here with any presumptuous proffer of information or instruction, either to practical or to theoretical farmers; but that I have come in simple deference to the repeated solicitations of friends, and because I have never learned that great art which the fairer portion of my audience understand how to prize and how to practise, when teased by the importunity of admiring suitors, — the art of saying *no!*

Seriously, my friends, I am here with a deep sense of my own insufficiency for these things, and with a full consciousness that there are hundreds around me to whom I might far better offer myself as a scholar, than as a teacher, upon any subject connected with the cultivation of the soil.

And yet, being here, and the responsibility for my presence being thus fairly rested upon other shoulders, I do not intend to shrink from the legitimate service of the occasion. Having once put my hand to the plough, I am not disposed "to look back," but shall proceed to break up such a furrow as I can, — to turn over as large a slice as I am able, — in some corner or other of the wide field of agricultural discussion. Before entering, however, upon the graver topics of the day, let me give expression to the emotions of pleasure with which I have always witnessed these Farmers' Festivals, as often as I have had an opportunity of attending them. They seem to me to come nearer to fulfilling the true idea of republican holidays, than any which our country has hitherto afforded. I know not how much they may do for the great interest which they are primarily designed to promote. It might not be easy to measure their precise effect in improving the cultivation, or enlarging the yield, of the soil, — though, even as to these ends, their influence, I am persuaded, is by no means inconsiderable. No one, indeed, can doubt, that for spreading information, for exciting and directing inquiry, for encouraging experiment, for stimulating emulation, and for exhibiting the practical and beneficial results of them all, such occasions furnish means and opportunities which could be supplied in no other way; and I venture to say, that there is not a farmer before me at this moment, who, if he should be rebuked on his return by some stay-at-home neighbor or by some over-anxious spouse, as having lost a day in attending the Cattle Show, would not confidently reply, that, instead of losing one day, he had gained ten, in the new ideas and fresh incentives which he had brought back for his future efforts.

But, however this may be, the influence of such occasions in other ways is even more appreciable. Their influence in the cultivation of good feelings and good fellowship among the friends of agriculture, and of labor generally, in different parts of the State and of the nation; their efficacy in sowing the seeds and increasing the harvest of mutual acquaintance, mutual regard, mutual respect, among all, of all classes, sexes, and occupations, who attend them; their annual operation in garnering up in the hearts of each one of us a seasonable supply of good-will and friendly sentiment towards each other, against the day when personal competitions or political conflicts shall come round to bring blight and mildew to so many of the nobler feelings of the soul, and to threaten starvation and famine to the whole better part of our nature, — these are among the results of such festivals as this, which must ever commend them to the regard of every Christian philanthropist.

You are here, my friends, from all quarters of the Old Colony, and from many other parts of the Commonwealth and of the country, from all pursuits and professions and political parties, to join hands and hearts in furtherance of the great industrial interests of the people. Some of you are here as practical producers, proud to display the results of your own labor and skill in the field or the dairy; and some of you have come as amateurs, gratified to behold the successes and achievements of your neighbors or friends. And we have all come as *consumers*, whether of our own or of other people's produce; and we all rejoice in the assurances and evidences which such occasions afford, that it will not be the fault of the ignorance or the idleness of man, if an abundance of the best food shall ever be wanting to ourselves or our children. But we have all come, too, I trust and believe, in no vain and arrogant reliance on human industry or human science for our daily bread, but with hearts grateful towards Heaven for the gracious promise that seed-time and harvest shall never fail, and for the great providential agencies to which we primarily owe whatever of agricultural success we have enjoyed or witnessed.

For, indeed, if there be any thing calculated to inspire a spirit of devout dependence and gratitude in the heart of man, it is the course of nature as contemplated in the operations of the husbandman. There are at least two things which a farmer can never do without, — the sun and the shower. No industry, no science, can supply their place. For almost every thing else there may be some sort of substitute contrived. But who can contrive a substitute for a day's sunshine, or even for an hour's rain? What artificial irrigation could prevent or mitigate the consequences of a midsummer's drought? What mechanical arrangement of stoves, what chemical evolution of heat, could stay the ravages of an early frost? How impotent is the arm of man, in presence of agencies like these, blighting in a week, or even nipping in a night, the whole result of a year of toil! We may invent curious implements and marvellous machines to save our own labor; but we can invent nothing which shall dispense with the blessing of God. Man may plough, man may plant; but man cannot give the increase. The great indispensable machinery of agriculture must ever be the "Mécanique Céleste," that sublime and stupendous system of suns and spheres and rolling orbs, moving on in serene and solemn majesty above us, and —

> "For ever singing, as they shine,
> The hand that made us is Divine."

And now, Mr. President and Gentlemen, I am here for no rhetorical display. I shall attempt nothing of the poetry or romance of agriculture. But I desire to invite your attention to a few plain and practical considerations, which have struck me as not unimportant or uninteresting in themselves, and as not inappropriate to an occasion of this sort.

Few things have been more noticeable, and few things, I am sure, more gratifying to us all, than the increased interest which has been lately manifested in many parts of the Union, and more especially in our own Commonwealth, in the honored cause for which you are associated. We have all witnessed with no ordinary satisfaction the efforts which have been made, and which have been so successfully made, to awaken the public mind to a deeper sense of the importance and dignity of agricultural pursuits. We have all rejoiced to find some of our ablest and most accomplished minds devoting themselves to subjects connected with the cultivation of land, the improvement of stock, the scientific analysis of soils and of plants, and the preservation and propagation of fruit-trees and forest-trees. The best wishes and the best hopes of us all have attended the local and the national conventions which have been held on the subject during the past year; and we have hailed with peculiar pleasure the establishment and organization of a Board of Agriculture, under the auspices of our own Commonwealth.

I think we shall acknowledge, however, that it is of the highest importance, at such a moment, that we should have some correct and exact ideas as to what is to be done, and as to what can be accomplished, in this behalf; that we should take a careful survey of the actual condition of American agriculture and of the real wants of the American farmer; so that we may propose to ourselves some definite, practical, and practicable ends, and so that our efforts may terminate in something better than vague promises, exaggerated estimates, and false expectations. We have been accustomed, of late years, to hear from some quarters of the country, and from some parts of the community, language of this sort: — Agriculture is a neglected interest. Government does nothing for it. Legislators, State and National, can find time and can find inducements for promoting and for protecting every other employment and occupation of the people. They can do every thing for commerce. They can do every thing for the fisheries. They can do every thing for manufactures and the mechanic arts. But the farmers can find nobody to do or to say any thing in their behalf.

Now, I will not stop to inquire directly how far this language is reasonable or just, either towards our State or National Governments. Nor will I do more than suggest, in this connection, that, if there has been any wrong of this kind, whether of omission or of commission, the redress has always been within the reach of the injured parties; the farmers having always been a great majority in the nation at large, embracing, it is estimated, "more than three-fourths of the population," and having thus had it always in their power to control the action of the Government at any time, through the simple agency of the elective franchise.

But taking it for granted, for a moment, that the allegation has been well laid, that the grievance has been real, that an interposition has at last been successfully made, and that the farmers are henceforth about to have their own way in the affairs of the country, I am disposed to ask some such questions as these:—What can Government do for American agriculture? What can it do for the interests and welfare of the farmers? What could it ever have done? What has it done or left undone hitherto?

I do not state these questions as distinct propositions, to be distinctly and formally treated in the order in which they have been stated, like the heads of an old-fashioned sermon, but as presenting the details of a general inquiry which I desire to institute, and, as far as possible, within the reasonable limits of such a discourse, to answer.

And here, at the outset, let me remark, that it is not altogether easy or practicable to treat the agricultural interests of the United States as a single idea, and to include them all as the subject of a common discussion. When we speak of British agriculture or of European agriculture, we have in our minds a homogeneous subject. But the vast territorial extent of our country, and its varied soils and climates and productions, prevent altogether that perfect unity and identity of interest which are found among the tillers of the earth in other lands. The planting interests of the Southern States present, I need not say, a totally different subject of discussion from the farming interests of the Northern and Western States. The character of the labor by which the great crops of the South are raised, and the purposes to which they are applied, make them an obvious exception to the general subject of American agriculture, or, at any rate, so distinct a branch of it as requires a distinct and separate consideration.

I intend, then, in these remarks, to confine myself to the agriculture which is carried on by the hands of freemen, and which is generally occupied in the production of food.

And in reference to American agriculture, as thus understood, I begin by asserting that Government can do little or nothing for its protection, in the sense in which the term "protection" is employed in such connections, by any direct means; and that, even were what is called "the Protecting System," the established policy of the country, it would be impossible to apply it to any considerable extent, directly and immediately, to agriculture.

The protection of agriculture is an idea plainly applicable to countries in which food cannot be produced in sufficient quantities to meet the wants of the population, or in which it cannot be produced at all, except at a higher cost than that at which it could be procured from other sources of supply. It supposes a competition, actual, or at least possible, in our own markets with the products of our own fields. It is a protection against something, and that something is obviously foreign importation.

Great Britain may be in a condition to protect her agriculture. And she did so in earnest, and most effectively, for a long series of years, by a systematic arrangement of prohibitory duties or sliding scales. She may now find it more consistent with her general welfare, — more for her advantage, in view of her manufacturing and commercial interests, — more for the improvement of her whole condition, to relax or abandon this system for a time or altogether. But this is a question with her of policy, and not of power. Nobody doubts that the state of British agriculture, the relation of production to population, the proportion of supply to demand, render it susceptible of this sort of governmental protection. And so it may be, and so it is, with other countries of the Old World, and perhaps of the New.

But what could prohibitory duties or sliding scales, applied to agricultural productions, accomplish for the American farmer? Is there any scarcity of food among us, inviting supplies from abroad? Can food be raised in other regions, and imported into our country, at lower rates than those at which we can raise it for ourselves? Do any foreign products of the soil enter into injurious competition with our own products in the American market? There may be a little flax-seed, a little coarse wool, or a few hides, brought here from South America or the East Indies; and now and then, during the prevalence of a mysterious blight, our provincial neighbors may supply us with a few potatoes, or even with a little wheat. But these are exceptional cases, entirely capable of explanation, if they were important enough to justify the consumption of time which such an explanation would involve.

The great peculiarity in the condition of the United States is, I need not say, its immense and immeasurable agricultural resources. Our boundless extent of fertile land, and the hardly more than nominal price at which it may be purchased, have settled the question for a thousand years, if not for ever, that, unless in some extraordinary emergency of famine or of civil war, our farmers will have the undisputed control of our own markets, without the aid of prohibitory duties or protective tariffs. It may be said to be with our lands, as it certainly is with our liberties: the condition of both may be described by the striking couplet of Dryden:—

> "Our only grievance is excess of ease,
> Freedom our pain, and plenty our disease."

Other Governments can do much more for political liberty than our Government can do, because there is so much more of this sort in other countries left to be done. We have a noble system of independence and freedom, already established and secured to us by the toil and treasure and blood of our fathers. We of this generation may say with the glorious apostle: "With a great price purchased they this freedom; but we were born free." The most, therefore, that any American Government can do now is to maintain, uphold, and administer, according to the true spirit and intent of those who acquired it, the ample patrimony of freedom which has been bequeathed to us. God grant that there may never be wanting to us rulers capable of doing so!

And now, my friends, Nature — I should rather say, a kind Providence — has done for our agricultural condition very much what the wisdom and valor of our fathers have effected for our political condition. It has given us a vast extent of virgin soil, susceptible of every variety of culture, and capable of yielding food for countless millions beyond our present population. It is ours to occupy, to enjoy, to improve and preserve it; and no protective systems are necessary to secure a market for as much of its produce as we, and our children, and our children's children for a hundred generations, can eat. Government can thus do nothing, nothing whatever, in the way of direct and immediate protection to American agriculture. And when it is said, therefore, that our legislators can protect commerce, can protect manufactures, can find time to look after all the interests of the merchant, the mechanic, the artisan, the navigator, and the fisherman, but can find no time to look after the interests of the farmer, — let it not be forgotten that such protection as may be afforded to commerce and manufactures, through the aid of a revenue system, is, from the nature of things, impracticable and impossible for agriculture. Let it not be forgotten, that, as to the great mass of human food which our soil supplies, we have a natural and perpetual monopoly in our own markets for as much as we can any way furnish mouths to consume or money to pay for. The ability to consume, in a word, pecuniary or physical, is the only limit to the demand for agricultural produce among ourselves; and this ability can by no possibility be affected by any legislative measures directed to the immediate promotion or protection of agriculture.

And here let me suggest a distinction, which, though often lost sight of, is, in this country at least, a real distinction, and not unworthy of serious attention: I mean the distinction between the promotion of agriculture, and the promotion of the immediate interests of those engaged in it. The promotion of agriculture looks obviously to an extended and an improved cultivation of the soil, to the introduction of better processes and better implements of agricultural labor, and to the consequent production of larger crops and more luxuriant harvests. But would such results be necessarily for the immediate benefit of the great body of American farmers? Would their condition, as individuals or as an aggregate class, be improved, — would their crops be enhanced in price, or stand a chance of commanding a convenient sale at any price, if the number of farmers were multiplied, if the breadth of land under cultivation were extended, and if, by the aid of greater science, of new manures, new machines, and new modes of culture, each one of them could double the yield of every acre of his land? Is it not obvious, that, unless new and adequate markets were simultaneously opened, the only consequence would be a still greater overplus of production, a still greater diminution of agricultural produce, and a still greater depression of the individual prosperity and welfare of the farmers?

The result of both the considerations which I have thus far suggested is the same. The great agricultural want of our country is the want of consumers and not of producers, of mouths and not of hands, of markets and not of crops. And this is a want which no government protection, like that which has been, or may be, afforded to manufactures or to commerce, can possibly supply. On the contrary, that sort of protection would only increase the difficulty, and aggravate the disease.

Indeed, the policy of our Government, in one particular at least, has already tended greatly to this result: I mean its *Public Land Policy.* Who can say that Government has done nothing for the protection of agriculture, who contemplates, for an instant, the course and consequences of this gigantic system? Consider the expenditure of care and of money, at which our vast territorial possessions have been acquired! Consider the expensive negociations, and the still more expensive wars, by which they have been purchased or conquered from foreign nations or from the Indian tribes! Consider the complicated and costly machinery of their survey and sale, and the systematic provisions which have been made for securing to every settler that first great want of an independent farmer, — a perfect title to his land! And then consider the almost nominal price at which any number of acres may be purchased!

I would not question the wisdom of this policy, for the purposes for which it was designed. It was designed to effect an early settlement and civilization of the great West; and its wisdom is justified by the existence, at so early a period after its adoption, of so many populous and prosperous States, in regions which were, seemingly but yesterday, the abodes of wild beasts or wilder men. We hail those new and noble States, as they successively and rapidly advance to maturity, as the proudest products of our land, and welcome them to the privileges and the glories of a Union which we pray may be perpetual.

The influences of this policy, in some other ways, may have been of a more doubtful character. But who can say that the American Government has done nothing for agriculture, with such a policy, so long and systematically pursued, before his eyes? What greater bounty could be contrived for the multiplication of farmers, and for the extended cultivation of the soil, than the standing offer of the best land in the world, with its title guaranteed by the strong arm of the nation, and its muniments deposited in the iron safes of the Government, at a dollar and a quarter an acre? — unless, indeed, it be found in the absolute gift of a homestead to every settler for two or three years, or in the "vote yourself a farm," or "land for the landless," projects of the present day. What has the Government ever done for commerce or for manufactures, which can compare with this great *bonus* to agriculture? Nay, what has the Government ever done, or ever been able to do, to counteract the constant drain upon commercial and manufacturing labor which this system has created?

No one, I suppose, can doubt that one of the great obstacles in the way of establishing and maintaining a manufacturing system, and of building up the mechanic arts, in these Eastern States, has been the constant inducement and temptation to leave home and go off to the West, which have been held out, in the fertility and cheapness of the Western lands, to the young men and young women, whose hands were essential to the loom, the spindle, the lapstone or the anvil. The absolute necessity of counteracting these inducements and temptations, by an increased rate of wages at home, has materially aggravated one of the greatest difficulties which we have encountered, in the way of a successful competition with the manufacturers of the old world. The influence of the luxuriant prairies and rich bottoms of Illinois, and Indiana, and Iowa, and Wisconsin, and the rest, has been similar to that of the placers and gold mines of California at the present moment; and, though less in degree, has been far more steady and durable than that is likely to be. Our young men and young women will not be long in learning, that there are more profitable diggings, in the long run, on this side of the Rocky Mountains than on the other. They will not be long in appreciating the philosophy of the cock, in the old fable of Æsop, who discovered that corn was a more reliable treasure than jewels. They will not be long in realizing, that even golden carrots may be a more certain crop than *carats* of gold. They will soon understand the wisdom of Franklin, in his conclusion of one of the numbers of the "Busy Body," — a little series of essays published by him in Philadelphia in 1729, and which, though among his earliest compositions, are replete with the wit and shrewdness and sterling common sense which characterized his maturer productions.

"I shall conclude," said he, "with the words of my discreet friend, *Agricola*, of Chester County, when he gave his son a good plantation, — 'My son, I give thee now a valuable parcel of land. I assure thee I have found a considerable quantity of gold by digging there: thee may'st do the same; but thee must carefully observe this, — never to dig more than plough-deep.'"

The temptations of good land will last longer than those of gold mines. There is a love for acres. There is a charm in independent proprietorship. There is health, and happiness, and a sense of freedom, in rural life and rural labor. There is a proud consciousness of virtue, and of worth, and of self-reliance, in the breast of the honest and industrious farmer, like that to which the simple shepherd of Shakspeare gave utterance, when reproached by the clown with a want of courtly manners:—

"Sir, I am a true laborer. I earn that I eat, get that I wear; owe no man hate, envy no man's happiness; glad of other men's good, content with my harm; and the greatest of my pride is to see my ewes graze, and my lambs suck."

Feelings and instincts like these, to which no bosom is a stranger, will outweigh and outlast the temptations of the richest placers of the Pacific, and will create a yearning towards the broad fields and noble forests of the great West, in the hearts of our enterprising young men and young women, as long as a single township or a single quarter section shall remain unsold or unsettled. That whole vast domain will thus continue to operate in the future, as it has operated in the past, as a continual government bounty upon the multiplication of farmers, and the extension of agriculture.

And now, having said thus much, and the limits of this address will not allow me to say more, both in regard to what Government cannot do for American agriculture, and also as to what it actually has done in the past, I come to a brief consideration of what it can do, and what it ought to do, in the future.

In the first place, it can adopt systematic, comprehensive, and permanent measures for ascertaining from year to year, or certainly from census to census, the actual condition of our country in relation to agriculture, the quantity of land under cultivation, the proportion of cultivated land devoted to the production of different articles of food, the relation of production to population in the various States and in the country at large, the comparative productiveness of the same crops in different parts of the Union and under different modes of culture, and generally whatever details may be included in a complete statistical account of American agriculture.

Our commercial and navigating statistics are already provided for, as incidental to our revenue-system. We need similar returns both of our agriculture and our manufactures; and I should not be sorry to have them committed to a common bureau.

One of the brief sayings, which have given a name and a perpetual fame to the Seven Wise Men of Ancient Greece, is the simple precept, "Know thyself." And a celebrated Latin poet has not been willing to regard it as a mere saying of human origin, but has emphatically declared that it descended from heaven.

It was a saying addressed to individual man, and undoubtedly contemplated that self-examination, that searching of the heart, which is a duty of higher than human authority, and which is essential to all moral or spiritual improvement. But it is a doctrine as applicable to the outer as to the inner man, and as essential to the progress and improvement of nations as of individuals. And this country, beyond all other countries, needs to know itself, to understand its own condition, to watch closely its own progress, to keep the *run* of it, as we may well say, for it is always on the run, advancing and going ahead with a rapidity never before witnessed, or dreamed of. More especially should the industry of our country know itself, and realize its own condition and circumstances. American labor, in all its branches, should have a map, on which it may behold its own aggregate position, and its own individual relations.

and by which it may be enabled to see what obstructions and interferences are in the way of its prosperous progress; to see particularly where it obstructs itself, by pressing into departments already too crowded, and where it may obtain relief and elbow-room in departments not yet occupied. American agriculture, above all, should be able to look itself fairly in the face, as in a mirror, through the medium of the most detailed and exact periodical surveys, that it may discover seasonably any symptoms of over-action or of under-action, if there be any; and that it may run no risk of expending and wasting its energies in unprofitable toils.

In the next place, Government, State and National, can encourage agricultural science, and promote agricultural education.

This subject has been so nearly exhausted, during the last year or two, by President Hitchcock's report to our own Legislature, by Dr. Lee's reports to the Patent Office at Washington, and by the lectures and addresses in which it has been treated in all parts of the country, that I propose to notice it very briefly.

Undoubtedly the noble system of common school education, which is already in existence among us, and for which we can never be too grateful to our Puritan Fathers, is itself no small aid to the cause of agriculture. The farmers, and the farmers' children, enjoy their full share of its benefits. It furnishes that original sub-soil ploughing to the youthful mind which is essential to the success of whatever other culture it may be destined to undergo. There is no education, after all, which can take the place of reading, writing, and keeping accounts; and the young man who is master of these elemental arts, and whose eye has been sharpened by observation, and his mind trained to reflection, and his heart disciplined to a sense of moral and religious responsibility, — and these are the great ends and the great achievements of our common schools, — will not go forth to the work of his life, whether it be manual or mental, whether of the loom or the anvil, of the pen or the plough, without the real, indispensable requisites for success. The great secret and solution of the wonderful advance which has been witnessed of late years, in all the useful arts, has been the union of the thinking mind and the working hand in the same person. Heretofore, for long ages, they have been everywhere separated. One set of men have done the thinking, and another set of men have done the working. The land has been tilled, the loom has been tended, the hammer and the hoe have been wielded, by slaves, or by men hardly more intelligent or independent than their brute yoke-fellows. In other countries, to a considerable extent, and even in our own, so far as one region and one race are concerned, this separation still exists. But a great change has been brought about by the gradual progress of free institutions; and, in the Free States of our own country especially, we see a complete combination of the working hand and the thinking mind, of the strong arm and the intelligent soul, in the same human frame. This has been the glorious result of our common school system, the cost of which, great as it has been and still is, has been remunerated a thousand fold, even in a mere pecuniary way, by the improvements, inventions, discoveries, and savings of all sorts, which have been made by educated labor in all the varied departments of human industry. It is now everywhere seen and admitted, that the most expensive labor which can be employed is *ignorant* labor; and, fortunately, there is very little of it left in the American market.

But, while the great substratum of all education for all pursuits is abundantly and admirably supplied by our common schools, no one can fail to perceive, or hesitate to admit, the advantages which may accrue from something of a more specific and supplementary instruction for those to whom the care and culture of the American soil is to be committed. The earth beneath us has been too long regarded and treated as something incapable of being injured by any thing short of a natural convulsion, or a providential cataclysm. We have been so long accustomed to dig it, and ditch it, and drain it, and hoe it, and rake it, and harrow it, and trample it under our feet, and plough long furrows in its back; and have so long found it repaying such treatment by larger and larger measures of endurance, generosity, and beneficence, — that we have been ready to regard it as absolutely insensible to injury. Because our chains and stakes have exhibited from year to year the same superficial measurements, we have flattered ourselves that our farms were undergoing no detriment or diminution. We have remembered the maxim of the law, "He who owns the soil owns it to the sky," and have been careful to let nothing interfere with our air or daylight; but we have omitted to look below the surface, and to discover and provide against the robbery which has been annually perpetrated, by day and by night, upon its most valuable ingredients and elements.

The discovery has at last been made, the danger has been revealed, the alarm has been sounded; and if Government can provide bounties for the destruction of the wolves and bears and foxes, which threaten our flocks, our herds, and our hen-roosts, I see not how it can withhold some seasonable provision against the far more frequent and more disastrous depredations by which our soil is despoiled of its treasures, through the want of science and skill on the part of those who till it. These depredations are none the less treacherous, or the less formidable, I need not say, for being carried on in no malicious spirit, and by no hostile hands. The worst robberies, of every sort, moral or pecuniary, of character, of property, or of opportunity, are those which a man commits upon himself. It is due to ourselves, it is due even more to our children, that the national soil should not be impaired by our ignorance or our neglect. It is a great trust-estate, of which each generation is entitled only to the use, and for the strip and waste of which the grand Proprietor of the Universe will hold us to account.

Whether the promotion of agricultural education shall be undertaken through systematic courses of scientific lectures, or by agricultural schools and colleges, with experimental farms attached to them, or by the preparation and distribution of agricultural tracts and treatises, or by all combined, it is for the farmers to say. What they say will not fail to be rightly and effectively said. With them words will be things; for no Government will venture to resist their deliberate and united appeals.

But let not the farmers, or the friends of the farmers, deceive themselves. When all that can be desired in this way shall have been accomplished; when Government shall have done its whole duty in regard to agricultural statistics and agricultural science; when the products of every State and of every district in the Union shall have been put in the way of exact and periodical ascertainment; when the American soil shall have been everywhere analyzed, and when those who till it shall have been everywhere instructed in its peculiar adaptations, and its peculiar properties, and its peculiar wants; when the whole vegetable and animal and mineral kingdoms shall have been raked and ransacked for the cheapest and most accessible and most effective fertilizers; when some safe and convenient mode shall have been contrived (according to the late suggestion of Lord Palmerston in England) for turning back the drains and gutters and common sewers of our great cities and towns upon our farms and gardens, instead of allowing them to run waste to the sea, breeding pestilence as they flow, "the country thus purifying the towns, and the towns fertilizing the country;" when the great doctrine of modern science shall be practically recognized and applied, that there is no waste in the physical universe, nothing in excess, nothing useless, from the bone which the dog growls over at our door, to the dung of the sea-fowl, for which the nations of the earth are contending, on the most distant and desolate island, but that

> "Nature never lends
> The smallest scruple of her excellence,
> But, like a thrifty goddess, she determines
> Herself the glory of a creditor,
> Both thanks and use;"—

still, still, the great want of American agriculture will remain, — that want which I have alluded to, in the opening of this address, and to which I recur once more, for a few moments, in its conclusion, — the want of adequate markets for the sale of its produce. Nay, the want will only have been increased and aggravated by the greater fertility of our fields, and the greater abundance of our harvests.

Now, it is obvious, that these markets are either to be supplied at home or abroad.

And I am not one of those, if any there be, who are disposed to disparage the value of a foreign market for any thing for which we can find one. It is clearly the duty of our Government to make arrangements in every way in its power by wise negotiations and just systems of reciprocity, for the introduction into foreign countries of the largest possible amount of our surplus provisions and breadstuffs. Such arrangements, however, are clearly commercial arrangements; and I refer to them merely as an illustration, that what may seem to be done by our legislators only for the benefit of commerce, may really result in the most important aid and advantage to agriculture.

I cannot pass from this topic, however, without the expression of an opinion, that the idea of an adequate foreign market for our agricultural surplus has proved, and will still prove, utterly fallacious and delusive. There is at least one principle, in this connection, which may be considered as settled by the whole current of experience, and by all the deductions and dictates of reason and common sense. No large or considerable kingdom or country will ever be habitually dependent on the soil of other countries for the food of its inhabitants. Why, where would be the power of Great Britain, were she compelled to look abroad for the daily bread of her people? What a mockery would be her boasted dominion over the seas! What a farce her world-encircling chain of colonial possessions and military posts! With what face would she venture to interfere with our fishing-grounds, or even to maintain her own, were she liable to be starved out at any moment by our embargoes! We should soon learn how to bring her to terms, as her own parliaments have so often brought her monarchs to terms, by a simple refusal of supplies, a simple stopping of rations.

I never think, Mr. President, of this dream of some of our American farmers, that they are to raise food for all the world, without associating it with the dream of Joseph of old, or rather with his two successive dreams, as related to his brethren, and recorded in Holy Writ: —

"Hear, I pray you," said he, "this dream which I have dreamed: For, behold, we were binding sheaves in the field, and lo! my sheaf arose, and also stood upright; and, behold, your sheaves stood round about, and made obeisance to my sheaf. And his brethren said to him, Shalt thou indeed reign over us? Or shalt thou indeed have dominion over us?"

"And he dreamed yet another dream, and told it to his brethren, and said: Behold, I have dreamed a dream more; and, behold, the sun, and the moon, and the eleven stars, made obeisance to me."

Sir, the one of these dreams is as likely to be fulfilled in our favor as the other. We may as well hope that the constellations of the other hemisphere will stoop to make obeisance to our constellation, and that the kings and queens of the earth will bend and do homage to our republic, as that the sheaves of other lands will stand round about and make obeisance to our sheaf, and the agriculture of the world acknowledge its dependence upon our agriculture.

Indeed, the fulfilment of the one dream, as I have already suggested, would speedily involve the fulfilment of the other. No great nation can ever maintain its political independence, except by sufferance and courtesy, when it has become absolutely dependent on another nation for its food. As to Great Britain, moreover, to whom our farmers have always been pointed for their most hopeful market, and to whom, I doubt not, they may always look confidently for an occasional demand for some varieties of agricultural produce, it is an admitted fact that she can feed herself, as it is, in all ordinary seasons; and when she shall have brought all her reserve land into cultivation, and reclaimed all her swamps and bogs and marshes, and established a better state of things for poor Ireland, and applied the modern modes of systematic, scientific culture to the whole soil of the United Kingdom, she may defy the farmers of the world. The whole notion of John Bull's submitting to be fed or foddered at our rack and out of our manger, is as visionary as that of Brother Jonathan's putting his neck back again under the old British yoke.

Nature herself, indeed, presents an obstacle which settles the question for ever. It has been calculated by the late lamented Mr. Porter, in his Progress of the British Nation (a work of standard authority), that "to supply the United Kingdom with the simple article of wheat would call for the employment of more than twice the amount of shipping which now annually enters our ports;" and that "to bring to our shores every article of agricultural produce in the abundance we now enjoy, would probably give constant occupation to the mercantile navy of the whole world."

The sum of the whole matter is this: American agriculture must look at home for its great market. It must look to consumers upon its own soil and at its own doors for its only sufficient and its all-sufficient demand. The natural and rapid increase of population among ourselves, and from the native stock, will do something for it. The thronging multitudes of emigrants, who are landed daily on our shores, will do something for it. If we cannot carry over our corn to the hungry millions of Europe, we can bring the hungry millions of Europe over to take for themselves from our granaries. This is the necessary course of things; and it is to be recognized and provided for,—not resisted, not complained of, but regulated and accepted cheerfully, as our part and lot in the dispensation of Providence. Our colonial fathers and mothers were pilgrims and exiles; and though we may look for no second May-flower, and no second Plymouth Rock, there are honest and heroic hearts beating beneath many a tattered frock or weather-beaten jacket from the Emerald Isle or the German Empire, which demand and deserve our sympathy and succor; and it would be a dishonor to the memory of our fathers, if we, their civilized descendants, should be found holding out a less hospitable reception to the homeless exile of the present day, than they received even from the poor untutored Indian, whom they were destined so sadly to displace and exterminate, when he cried to them, "Welcome, Englishmen!"

But something more than the increase of population, whether by multiplication at home or by immigration from abroad, is necessary for the relief and just remuneration of

American agriculture. Indeed (as I have already suggested), if these throngs of emigrants, and if so many of the young men and women of our own stock, are to swarm over at once to our Western lands, and enter forthwith upon a life of agricultural production, they will only increase and aggravate the difficulties under which our farmers already labor. Instead of population gaining upon food, food will still go on gaining upon population; instead of mouths waiting for bread, we shall perpetuate the spectacle of bread waiting, and waiting in vain, for mouths.

In one word, there must be a division and distribution of labor in our country, to a much greater extent than exists at present, in order that agricultural industry may receive its just rewards. There must be more, and more numerous, separate classes of consumers, distinct from the producers, in order that food may command a fair price, and afford an adequate compensation and encouragement to the labor which is employed in raising it. Cheap food is a blessing not to be spoken lightly of; but the laborer is worthy of his hire, and it can never be the policy of any country to have food so cheap that it shall not pay for the raising, that it shall not pay something more than the mere cost of the raising. It can never be the policy of a free republican country like ours, where the most important rights and duties of Government are enjoyed and exercised by all men alike and equally, and where intelligence, education, and individual independence are essential to the maintenance of our liberties, to reduce either the profits of land or the wages of labor to the standard of a bare subsistence.

Farming is never destined to be a means of fortune-making, and we may all thank Heaven that it is so. If millionnaires and capitalists and speculators could make their cent per cent per annum by growing corn, we should soon see our land bought up for permanent investment for hirelings to till; and our little independent proprietors, cultivating their own acres, would be no longer the stay and staff of our republican institutions and our republican principles. God grant that the day may never come, when this country shall be without an independent rural population, owning no lord or master this side of Heaven; maintaining, in all their purity and freshness, those rural manners and rural habits which are the very salt and saving grace of our social and our political system. God grant that the day may never come, when some American Goldsmith shall paint our rural villages deserted, our rural virtues leaving the land:—

> "E'en now, methinks, as pondering here I stand,
> I see the rural virtues leave the land.
> Contented toil, and hospitable care,
> And kind, connubial tenderness, are there;
> And piety with wishes placed above,
> And steady loyalty, and faithful love."

But the farmer ought to have something more than a mere living price for his products. He ought to be able to lay up something to send a son to college, or to set up a daughter in house-keeping, or to support his wife and himself, and keep the wolf from the door, when sickness or old age shall put a stop to their daily toil. The true protection of agriculture, and the true promotion of the welfare of the individual farmer, are to be found, and can only be found, in building up the manufacturing and mechanic arts of our country, in creating a diversified industry, and in establishing more proportionate relations between the various departments of human labor. When this shall be accomplished, there will be less need of Government intervention for encouraging agricultural science and diffusing agricultural information. It will then cease to be recorded of our American agriculture, that "its two prominent features are its productiveness of crops, and its destructiveness of soil;" for it is the one of these features which leads directly to the other. It is the over-production of our agriculture which causes so much of careless and destructive cultivation. It is the superabundance of our aggregate harvests which occasions the meagreness of so many of our individual harvests. Who cares to make his farm yield double its present crop, when there is so precarious a market for what it yields already? Who can style him a benefactor who makes two blades of grass grow where only one grew before, when the result of such a process must be to diminish the chances of remuneration to the laborer, and when doubling the product is so likely to divide an already inadequate price?

And now, my friends, I am not about to violate the political neutrality of this occasion, by inquiring how this diversified industry, which is so necessary to the prosperity of the farmer, and to the promotion of agriculture, is to be brought about; whether by protective tariffs, or judicious tariffs, or moderate specific duties, or reasonable discrimination, or by ad-valorems and free trade. This question, though it never ought to have been permitted to enter into party politics, has practically become so identified with them, that it must be left to other occasions. But the necessity of a greater distribution of labor to the prosperity of all concerned in labor, and the especial need which the American farmer feels, at this moment, of more persons engaged in other pursuits, who may become purchasers and consumers of his produce, and the danger that the American soil will receive serious and permanent detriment from the careless, hand-to-mouth, cultivation, which such a state of things induces, — these are no party topics. They are great truths, which all must admit, and which all ought to lay to heart.

There is a letter of Dr. Franklin's, written in London on the 22d of April, 1771, to Humphry Marshall, a Pennsylvania Farmer, which contains as much practical wisdom as I ever remember to have found in the same compass, in relation to the prosperity of the American farmer. It is as applicable now as when it was written; and it ought to be printed in good legible type, and hung up in a frame in every farmer's house in the Union:—

"The Colonies," says he, "that produce provisions, grow very fast. But, of the countries that take off those provisions, some do not increase at all, as the European nations; and others, as the West India Colonies, not in the same proportion. So that, though the demand at present may be sufficient, it cannot long continue so. Every manufacturer encouraged in our country makes part of a market for provisions within ourselves, and saves so much money to the country as must otherwise be exported to pay for the manufactures he supplies. Here in England," he adds, "it is well known and understood, that, wherever a manufacture is established which employs a number of hands, it raises the value of lands in the neighboring country all around it, partly by the greater demand near at hand for the produce of the land, and partly from the plenty of money drawn by the manufacturers to their part of the country. It seems, therefore, the interest of all our farmers and owners of lands to encourage our young manufactures in preference to foreign ones, imported among us from distant countries."

In these golden words of Franklin, which could find no better illustration the world over than here, in presence of those to whose lands and to whose crops yonder mills and furnaces and machine-shops have given a value so far beyond any which they could otherwise have commanded, —if these golden words of Franklin, I say, could be impressed upon the heart and mind of every farmer in our land, there would be less complaint that our Government had found time to do every thing for manufactures and the mechanic arts, and had done nothing for agriculture; and it would be seen and understood, that whatever had been done for any one of the great interests of American labor had been done for all; and that all were bound up together for a common weal or a common woe, incapable of separation or opposition. There is nothing indeed more evident, and nothing more beautiful, than the harmony of all the great industrial interests in our Union. There may be jealousies and rivalries and oppositions between the farmers and the manufacturers and the merchants elsewhere, in the old, closely settled, and crowded populations of Europe; but there can be none reasonably, none rightfully, here. Nothing short of miraculous intervention, like that which watered the fleece of Gideon, while all the other fleeces were dry, can elevate one branch of industry, or one department of labor, at the expense of another. The highest prosperity of each is not only consistent with, but inseparable from, the highest prosperity of all. What is done for any is done for all; and all find their best encouragement and protection in the common welfare and prosperity of the whole community. We see, or ought to see, something of that mutual sympathy and succor among American laborers, of which so graphic a sketch is given by one of the prophets of Israel: "So the carpenter encouraged the goldsmith, and he that smootheth with the hammer him that smote the anvil. They helped every one his neighbor; and every one said to his brother, Be of good courage."

The greatest division of labor, the most complete and cordial union among laborers, — this is the true motto and maxim which our condition suggests and inculcates; and the American farmer should be the first to adopt and cherish it.

A word or two, Mr. President and gentlemen, and only a word or two, in conclusion. In all that I have said, I have spoken, as I proposed to speak, of American agriculture, so far as it is occupied in the production of food, and through the agency of free labor, in all parts of our wide-spread land. In looking at the agriculture of Massachusetts as a separate State, we find many of the circumstances, which characterize the agricultural condition of the country at large, reversed. There is no over-production of food, and no danger of any such over-production, for our own population within our own limits. On the contrary, it has been estimated that we are at this moment dependent on our sister States for more than three millions of bushels of breadstuffs, — being a full half of our whole consumption.

Now, so far as this fact may fairly betoken any bad cultivation on the part of our farmers; so far as, taken in connection with other facts, it indicates a deterioration of our soil, and a progressive disproportion between the acres in cultivation and the crops which they yield, — it is a fact deeply to be deplored, and which ought to furnish a serious warning to the Government and the people of the Commonwealth.

But, so far as it only indicates a greater division and distribution of labor within our own borders; so far as it is only the result of a gradual multiplication of mechanics and manufacturers among us, to consume the products, not only of our own husbandmen, but of those of other States, neighboring and remote, — it is a subject of positive and unqualified congratulation. For one, I never desire to see the day when Massachusetts shall feed herself. Nature has marked and quoted her for a different destiny. Her long line of indented sea-coast, stretching out around two noble capes, and bending in again along two noble bays, designates her unmistakably for a commercial and navigating State; and her countless fleets of coasters and fishing smacks and merchant-ships and whalers give ample attestation that she has not been blind to her vocation. Her numerous rivers and streams, with their abundant waterfalls, designate her hardly less distinctly as a manufacturing

State; and her sons, and her daughters too, are fast proving that they know how to fulfil this destiny also. A great agricultural State she was never made for. If she ever feeds herself, it will be by the decrease of her population, and not by the adequacy of her products. Her farmers will always find enough to occupy them. The perishable articles of daily consumption, which must be found at one's door, or not at all, must come always from them. Their milk, their garden-fruits and vegetables, their hay too, and their eggs and poultry, can hardly be interfered with injuriously, if at all, by any supplies from abroad, and can hardly be furnished in too large quantities at home. But the cereal grains, the beef and pork and mutton, and the butter and cheese of other States, are, I trust, to find a still increasing market in Massachusetts, in exchange for the products of her looms and anvils and lap-stones, and for the earnings of her commerce and fisheries. I would gladly see the United States independent of all foreign nations for all the necessaries of life, — clothing as well as food; but I do not desire to see the separate States independent of each other: first, because climate, soil, geographical position, and physical condition, designate them for different departments of industry, and their own highest prosperity will be subserved by following nature; and, second, because these mutual wants and mutual dependencies are among the strongest bonds of our blessed Union, and give the best guaranty that it shall endure for ever.

Let Massachusetts do all the farming she can; and all that she does, let her be sure to do well. Let her transmit no exhausted or impoverished soil to posterity. Let her exhibit to all the world what industry and energy and thrift and temperance and education and science can do, in overcoming the disadvantages and obstacles of a hard soil and a stern sky. Let her be a model State in agriculture, and in whatever else she undertakes. But let her not dream of feeding herself. For myself, I should feel as if either the days of the American Union were numbered, or certainly as if her own house were about to be left unto her desolate, if the time should ever come when the wheat of Pennsylvania and Maryland, and the pork of Ohio, and the beef and mutton of New York and Vermont, and the yellow corn of Virginia, and the rice of the Carolinas, could find no ready market for their sale, and no willing and watering mouths for their consumption, in the old Bay State. I delight to contemplate the various members of this vast republic, like members of a common family, not all alike, but with only such distinctions as become sisters; not selfishly and churlishly attempting to do every thing for themselves, or to interfere with each other's vocation, but pursuing their different destinies in a spirit of mutual kindness and mutual reliance; freely interchanging the products of their soil and of their skill in time of peace, and firmly interposing their united power for the common protection in time of war; bearing each other's burdens; supplying each other's wants; remembering each other's weaknesses; rejoicing in each other's prosperity; and all clustering with eager affection around the car of a common Liberty, — like the Hours in the exquisite fresco of Guido around the chariot of the Sun, — as it advances to scatter the shades of ignorance and oppression, and to spread light and freedom and happiness over the world!

Gentlemen, I can offer no better prayer to Heaven, either for human liberty or for human labor in all its branches, than that this spectacle of concord and harmony among the American States may be witnessed in still increasing beauty and perfection, as long as the Sun or the Hours shall roll on!

www.ingramcontent.com/pod-product-compliance
Lightning Source LLC
LaVergne TN
LVHW011124110826
845150LV00008B/2245